BIBLIOGRAPHIC RECORD TARGET

Graduate Library
University of Michigan

Preservation Office

Storage Number: ______________________

ACV1826
UL FMT B RT a BL m T/C DT 07/19/88 R/DT 07/19/88 CC STAT mm E/L 1
010: : |a 03023231
035/1: : |a (RLIN)MIUG86-B37540
035/2: : |a (CaOTULAS)160647320
040: : |c MnU |d MnU |d MiU
050/1:0 : |a QA462 |b .S74
100:1 : |a Spencer, William George.
245:00: |a Inventional geometry; |b a series of problems, intended to familiarize the pupil with geometrical conceptions, and to exercise his inventive faculty. |c By William George Spencer. With a prefatory note by Herbert Spencer.
260: : |a New York, |b D. Appleton and Company, |c 1877.
300/1: : |a 97 p. |c 16 cm.
490/1:0 : |a Science primers
650/1: 0: |a Geometry |x Study and teaching
700/1:1 : |a Spencer, Herbert, |d 1820-1903.

INVENTIONAL GEOMETRY:

A SERIES OF PROBLEMS,

INTENDED TO

FAMILIARIZE THE PUPIL WITH GEOMETRICAL CONCEPTIONS,
AND TO EXERCISE HIS INVENTIVE FACULTY.

BY

WILLIAM GEORGE SPENCER.

WITH A PREFATORY NOTE

BY HERBERT SPENCER.

NEW YORK:
D. APPLETON AND COMPANY,
549 & 551 BROADWAY.
1877.

PREFACE TO THE AMERICAN EDITION.

This little book, prepared by an experienced mathematical teacher for the use of his own pupils, is based upon the principle that the best and only true education is self-education. It introduces the beginner to geometry by putting him at work on problems which will not only thoroughly familiarize his mind with geometrical ideas, but will exercise, at the same time, his inventive and constructive faculties—a kind of mental practice of much importance, but generally neglected in our schools. These problems, which are simple at first and skillfully graded, the pupil is to solve, himself, without assistance. The author prepared no key to the work, considering that any such help in getting through it would defeat its purpose.

As this little book seems well suited to accompany the "Science Primers," that are now appearing from time to time, it has been gotten up in the same form, and is included among the American reprints of that elementary series.

The author of this volume of exercises was the father of Herbert Spencer, the eminent philosophical thinker, and whose valuable work on Education has been translated into nearly all the languages of Europe. He cordially commends the method of the "Inventional Geometry" from both observation and experience, as will be seen by the following letter:

NOTE FROM HERBERT SPENCER.

LONDON, *June* 3, 1876.

MESSRS. D. APPLETON & CO.: I am glad that you are about to republish, in the United States, my father's little work on "Inventional Geometry." Though it received but little notice when first issued here, recognition of its usefulness has been gradually spreading, and it has been adopted by some of the more rational science-teachers in schools. Several years ago I heard of its introduction at Rugby.

To its great efficiency, both as a means of producing interest in geometry and as a mental discipline, I can give personal testimony. I have seen it create in a class of boys so much enthusiasm that they looked forward to their geometry-lesson as a chief event in the week. And girls initiated in the system by my father have frequently begged of him for problems to solve during their holidays.

Though I did not myself pass through it—for I commenced mathematics with my uncle before this method had been elaborated by my father—yet I had experience of its effects in a higher division of geometry. When about fifteen, I was carried through the study of perspective entirely after this same method: my father giving me the successive problems in such order that I was enabled to solve every one of them, up to the most complex, without assistance.

Of course, the use of the method implies capacity in the teacher and real interest in the intellectual welfare of his pupils. But given the competent man, and he may produce in them a knowledge and an insight far beyond any that can be given by mechanical lesson-learning.

Very truly yours,

HERBERT SPENCER.

INTRODUCTION.

When it is considered that by geometry the architect constructs our buildings, the civil engineer our railways; that by a higher kind of geometry, the surveyor makes a map of a county or of a kingdom; that a geometry still higher is the foundation of the noble science of the astronomer, who by it not only determines the diameter of the globe he lives upon, but as well the sizes of the sun, moon, and planets, and their distances from us and from each other; when it is considered, also, that by this higher kind of geometry, with the assistance of a chart and a mariner's compass, the sailor navigates the ocean with success, and thus brings all nations into amicable intercourse—it will surely be allowed that its elements should be as accessible as possible.

Geometry may be divided into two parts—practical and theoretical: the practical bearing a similar relation to the theoretical that arithmetic does to algebra. And just as arithmetic is made to precede algebra, should practical geometry be made to precede theoretical geometry.

Arithmetic is not undervalued because it is inferior to algebra, nor ought practical geometry to be despised because theoretical geometry is the nobler of the two.

However excellent arithmetic may be as an instrument for strengthening the intellectual powers, geometry is far more so; for as it is easier to see the relation of surface to surface and of line to line, than of one number to another, so it is easier to induce a habit of reasoning by means of geometry than it is by means of arithmetic. If taught judiciously, the collateral advantages of practical geometry are not inconsiderable. Besides introducing to our notice, in their proper order, many of the terms of the physical sciences, it offers the most favorable means of comprehending those terms, and impressing them upon the memory. It educates the hand to dexterity and neatness, the eye to accuracy of perception, and the judgment to the appreciation of beautiful forms. These advantages alone claim for it a place in the education of all, not excepting that of women. Had practical geometry been taught as arithmetic is taught, its value would scarcely have required insisting on. But the didactic method hitherto used in teaching it does not exhibit its powers to advantage.

Any true geometrician who will teach practical geometry by definitions and questions thereon, will find that he can thus create a far greater interest in the science than he can by the usual course; and, on adhering to the plan, he will perceive that it brings into earlier activity that highly-valuable but much-neglected power, the power to invent. It is this fact that has induced the author to choose as a suitable name for it, the inventional method of teaching practical geometry.

He has diligently watched its effects on both sexes, and his experience enables him to say that its tendency is to lead the pupil to rely on his own resources, to systematize his discoveries in order that he may use them, and to gradually induce such a degree of self-reliance as enables him to prosecute his subsequent studies with satisfaction: especially if they should happen to be such studies as Euclid's "Elements," the use of the globes, or perspective.

A word or two as to using the definitions and questions. Whether they relate to the mensuration of solids, or surfaces, or of lines; whether they belong to common square measure, or to duodecimals; or whether they appertain to the canon of trigonometry; it is not the author's intention that the definitions should be learned by rote; but he recommends that the pupil should give an appropriate illustration of each as a proof that he understands it.

Again, instead of dictating to the pupil how to construct a geometrical figure—say a square—and letting him rest satisfied with being able to construct one from that dictation, the author has so organized these questions that by doing justice to each in its turn, the pupil finds that, when he comes to it, he can construct a square without aid.

The greater part of the questions accompanying the definitions require for their answers geometrical figures and diagrams, accurately constructed by means of a pair of compasses, a scale of equal parts, and a protractor, while others require a verbal answer merely. In order to place the pupil as much as possible in the state in which Nature places him, some questions have been asked that involve an impossibility.

Whenever a departure from the scientific order of the questions occurs, such departure has been preferred for the sake of allowing time for the pupil to solve some difficult problem; inasmuch as it tends far more to the formation of a self-reliant character; that the pupil should be allowed time to solve such difficult problem, than that he should be either hurried or assisted.

The inventive power grows best in the sunshine of encouragement. Its first shoots are tender. Upbraiding a pupil with his want of skill, acts like a frost upon them, and materially checks their growth. It is partly on account of the dormant state in which the inventive power is found in most persons, and partly that very young beginners may not feel intimidated, that the introductory questions have been made so very simple.

TO THE PUPIL.

WHEN it is found desirable to save time, omit copying the definitions; but when time can be spared, copy them into the trial-book, to impress the terms on the memory.

In constructing a figure that you know, use arcs if you prefer them; but, in all your attempts to solve a problem, prefer whole circles to arcs. Circles are suggestive, arcs are not.

Always have a reason for the method you adopt, although you may not be able to express it satisfactorily to another. Such, for example, as this: If from one end of a line, as a centre, I describe a circle of a certain size, and then from the other end of the line, as another centre, I describe another circle of the same size, the points where those circles intersect each other, if they intersect at all, must have the same relation to one end of such line which they have to the other.

The most improving method of entering the solutions is to show, in a first figure, all the circles in full by which you have arrived at the solution, and to draw a second figure in ink, without the circles.

It is not so much the problems which you are assisted in performing, as the problems you perform yourself, that will improve your talents and benefit your character. Refrain, then, from looking at the constructions invented by other persons—at least till you have discovered a construction of your own. The less assistance you seek the less you will require, and the less you will desire.

As the power to invent is ever varying in the same person, and as no two persons have that power equally, it is better not to be anxious about keeping pace with others. Indeed, all your efforts should be free from anxiety. Pleasurable efforts are the most effective. Be assured that no effort is lost, though at the time it may appear so. You may improve more while studying one problem that is rather intricate to you, than while performing several that are easy. Dwell upon what the immortal Newton said of his own habit of study. "I keep," says he, "the subject constantly before me, and wait till the first dawnings open by little and little into a full and clear light."

INVENTIONAL GEOMETRY.

The science of relative quantity, solid, superficial, and linear, is called Geometry, and the practical application of it, Mensuration. Thus we have mensuration of solids, mensuration of surfaces, and mensuration of lines; and to ascertain these quantities it is requisite that we should have dimensions.

The top, bottom, and sides of a solid body, as a cube,[1] are called its faces or surfaces,[2] and the edges of these surfaces are called lines.

The distance between the top and bottom of the cube is a dimension called the height, depth, or thickness of the cube; the distance between the left face and the right face is anoth-

[1] The most convenient form for illustration is that of the cubic inch, which is a solid, having equal rectangular surfaces.

[2] A surface is sometimes called a superficies.

er dimension, called the breadth or width ; and the distance between the front face and the back face is the third dimension, called the length of the cube.

Thus a cube is called a magnitude of three dimensions.

The three terms most commonly applied to the dimensions of a cube are length, breadth, and thickness.

1. Place a cube with one face flat on a table, and with another face toward you, and say which dimension you consider to be the thickness, which the breadth, and which the length.

2. Show to what objects the word *height* is more appropriate, and to what objects the word *depth*, and to what the word *thickness*.

As a surface has no thickness, it has two dimensions only, length and breadth. Thus a surface is called a magnitude of two dimensions.

3. Show how many faces a cube has.[1]

[1] The surfaces of a cube are considered to be plane surfaces.

When a surface is such, that a line placed anywhere upon it will rest wholly on that surface, such surface is said to be a plane surface.[1]

As a line has neither breadth nor thickness, it has one dimension only, that of length.

Thus a line is called a magnitude of one dimension.

4. Count how many lines are formed on a cube by the intersection of its six plane surfaces.

If that which has neither breadth, nor thickness, but length only, can be said to have any form, then a line is such, that if it were turned upon its extremities, each part of it would keep its own place in space.

We cannot with a pencil make a line on paper—we represent a line.

The boundaries or ends of a line are called points, and the intersection of two lines gives a point.

As a point has neither length, breadth, nor

[1] When the word *line* is used in these definitions and questions a straight line is always meant.

thickness, it is said to have no dimension. It has position only.

A point is therefore not a magnitude.

5. Name the number of points that are made by the intersection of the twelve lines of a cube.

We cannot with a pencil make a point on paper—we represent a point.

When any two straight lines meet together from any other two directions than those which are perfectly opposite, they are said to make an angle.

And the point where they meet is called the angular point.

Thus two lines that meet each other on a cube make an angle.

6. Represent on paper a rectilineal angle.

7. Can two lines meet together without being in the same plane?

8. Point out two lines on a cube that exist on the same surface, and yet do not make an angle.

9. Name the number of plane angles on

the six surfaces of a cube, and the number of angular points, and say why the angular points are fewer than the plane angles.

The meeting of two plane surfaces in a line —for example, the meeting of the wall of a room with the floor, or the meeting of two of the surfaces of a cube—is called a dihedral angle.[1]

10. Say how many dihedral angles a cube has.

The corner made by the meeting of three or more plane surfaces is called a solid angle.

11. Say how many solid angles there are in a cube.

When a surface is such that a line, when resting upon it in any direction, will be touched by it toward the middle of the line only, and not at both ends, such surface is called a convex surface.

12. Give an example of a convex surface.

When a surface is such that a line while resting upon it, in any direction, will be touched by

[1] Dihedral means two-surfaced.

it at the ends, and not toward the middle of the line, such surface is called a concave surface.

13. Give an example of a concave surface.

A simple curve is such, that on being turned on its extremities, every point along it will change its place in space; so that, in a simple curve, no three points are in a straight line.

14. Give an example of a simple curve.

Lines or curves grouped together by way of illustration, or for ornament, without regard to magnitude or surface, take the name of diagrams.

15. Give an example of a diagram.

When a surface [1] is spoken of with regard to its form and size, it takes the name of figure.

If the boundaries of a surface are straight lines, the figure is called a rectilinear figure, and each boundary is called a side.

Thus we have rectilinear figures of four sides, of five sides, of six sides, etc.

16. Make a few rectilinear figures.

[1] In the definitions and questions of this work, when the word *surface* is used, a plane surface is meant.

When a surface is inclosed by one curve, it is called a curvilinear figure, and the boundary is called its circumference.

17. Make a curvilinear figure with one curve for its boundary, and in it write its name, and around it the name of its boundary.

18. Make a curvilinear figure with more than one curve for its boundaries.

A figure bounded by a line and a curve, or by more lines and more curves than one, is called a mixed figure.

19. Make a mixed figure, having for its boundaries a line and a curve.

20. Make a mixed figure, having for its boundaries one line and two curves.

21. Make a mixed figure, having for its boundaries one curve and two lines.

When a figure has a boundary of such a form that all lines drawn from a certain point within it to that boundary are equal to one another, such figure is called a circle, and such point is called the centre of that circle; and the boundary is called the circumference of the circle, and the equal lines drawn from the centre to the circumference are called the radii of the circle.

22. Make four circles. On the first write its name. Around the outside of the second, write the name of the boundary. In the third, write against the centre its name. And between the centre and the circumference of the fourth circle, draw a few radii and write on each its name.

23. Can you place two circles to touch each other at a particular point?

24. Can you place three circles in a row, and let each circle touch the one next to it?

A part of the circumference of a circle is called an arc.

When the circumference of a circle is divided into two equal arcs, each arc is called a semi-circumference.

All arcs of circles which extend beyond a semi-circumference are called greater arcs.

All arcs of circles that are not so great as a semi-circumference are called less arcs.

A line that joins the extremities of an arc is called the chord of that arc.

When two radii connect together any two points in the circumference of a circle which are on exactly the opposite sides of the centre, they make a chord, which is called the diameter of the circle, and such diameter divides the circle into two equal segments,[1] which take the name of semicircles.

25. Make a circle, and in it draw two radii in such a position as to divide it into two equal parts, and write on each part its specific name.

All segments of a circle which occupy more than a semi-circle are called greater segments.

26. Make a greater segment, and on it write its name.

27. Make a greater segment, and on the outside of each of its boundaries write its name.

[1] The word *segment* means a piece cut off: thus we have segments of a line and segments of a sphere, as well as segments of a circle.

All segments of a circle that do not contain so much as a semi-circle are called less segments.

28. Make a less segment, and in it write its name.

29. Make a less segment, and on the outside of each of its boundaries write its name.

30. Can you cut from a circle more than one greater segment?

31. Can you cut from a circle more than one less segment?

32. Place two circles so that the circumference of each may rest upon the centre of the other, and show that the curved figure common to both circles consists of two segments, and may be called a double segment.

33. In how many ways can you divide a double segment into two equal and similar parts?

34. In how many ways can you divide a double segment into four equal and similar parts?

35. Can you make two angles with two lines?

When two lines are so placed as to make two angles, one of the lines is said to stand upon the other, and the angles they thus make are called adjacent angles.

36. Make two unequal adjacent angles with two lines.

When one line stands upon another line, in such a direction as to make the adjacent angles equal to one another, then each of these angles is called a right angle.

37. Make two equal adjacent angles, and in each angle write its proper name.

Either of the sides of a right angle is said to be perpendicular to the other; and the one to which the other is said to be perpendicular is called the base.

38. Make a right angle, and against the sides of the right angle write their respective names.

39. Can you make three angles with two lines?

40. Can you make four angles with two lines?

41. Can you make more than four angles with two lines?

42. Can you divide a line into two equal parts?

43. Can you divide an arc into two equal parts?

You have been told that figures bounded by lines are called linear figures.

44. Make a linear figure having the fewest boundaries possible, and in it write its name, and say why such figure claims that name.[1]

When a figure has for its boundaries three equal lines, it is called an equilateral triangle.[2]

45. Can you make an equilateral triangle?

46. Can you with three lines make two angles, three, four, five, six, seven, eight, nine, ten, eleven, twelve, thirteen?

47. Can you so place two equilateral triangles that one side of one of them may coincide with one side of the other?

48. Can you divide an equilateral triangle into two parts that shall be equal to each other and similar to each other?

[1] Triangles are also called trilaterals.

[2] Equilateral triangles are also called trigons.

49. Can you draw one line perpendicular to another line, from a point that is in the line but not in the middle of it?

The figure formed by two radii and an arc is called a sector.

When a circle is divided into four equal sectors, each of such sectors takes the name of quadrant.

50. Divide a circle into four equal sectors, and write upon each sector its specific name.

51. Make a set of quadrants, and write in each angle its specific name.

To compare sectors of different magnitudes with each other, geometricians have found it convenient to imagine every circle to be divided into three hundred and sixty equal sectors; and a sector consisting of the three hundred and sixtieth part of a circle, they have called a degree. An arc, therefore, of such a sector is an arc of a degree;[1] and the angle of such a sector is an angle of a degree.

[1] A degree of a circle is concisely marked thus (1°). Thirty degrees thus (30°). Thirty-five degrees thus (35°).

52. Make a set of quadrants, and write in each angle how many degrees it contains.

All angles greater or less than the angle of a quadrant are called oblique angles.

When an oblique angle is less than a quadrantal angle, that is less than a right angle, that is less than an angle of 90°, it is called an acute angle.

53. Make an acute angle.

When an oblique angle has more degrees in it than 90°, and less than 180°, it is called an obtuse angle.

54. Make an obtuse angle.

55. Make an acute-angled sector.

56. Make an obtuse-angled sector.

When a sector has an arc of 180°, the radii forming with each other one straight line, it has the same claim to be called a sector as it has to be called a segment, and yet it seldom takes the name of either, being generally called a semicircle.

57. Make three sectors, each containing 180°,

and write in each sector a different name, and yet an appropriate one.

A sector which has an arc greater than a semi-circumference is said to have a reëntrant angle.

58. Make a reëntrant-angled sector.

59. Say to which class of sectors the degree belongs.

You have halved a line, and you have halved an arc.

60. Can you divide a segment into two parts that shall be equal to each other, and similar to each other?

61. Can you divide a sector into two parts that shall be equal to each other, and similar to each other?

It is said by some, the circumference of a circle is 3 times its own diameter; by others, more accurate, that it is $3\frac{1}{7}$ times its own diameter.

62. Say how you would determine the ratio the circumference of a circle bears to its diam-

eter, and say also what you make the ratio to be.

You have divided a line, an arc, a segment, and a sector, into two equal parts.

63. Can you divide an angle into two equal parts?

When a triangle has two only of its sides of equal length it is called an isosceles triangle.

64. Make an isosceles triangle.

When a triangle has all its sides of different lengths it takes the name of scalene.

65. Make a scalene triangle.

When a triangle has one of its angles a right angle, it is called a right-angled triangle.

66. Make a right-angled triangle.

When a triangle has each of its angles less than a right angle, and all different in size, it is called a common acute-angled triangle.

67. Make a common acute-angled triangle.

When a triangle has one of its angles obtuse, it is called an obtuse-angled triangle.

68. Make an obtuse-angled triangle.

In describing the properties of a triangle it is not unusual to mark each angular point of the triangle with a letter.

Thus the accompanying triangle is called the triangle A B C, and the sides are called A B, B C, and A C, and the three angles are called A, B, C, or the angles C A B, A B C, A C B.

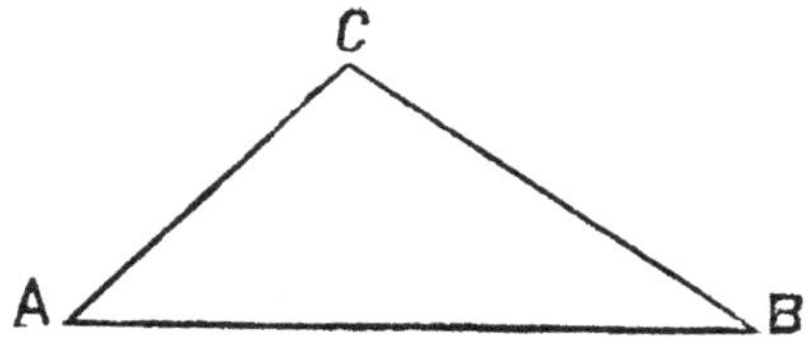

69. Can you make an isosceles triangle without using more than one circle?

When two lines do not meet either way, though produced ever so far, they are said to be parallel.[1]

70. Draw two parallel lines.

71. Can you draw one line parallel to another, and let the two be an inch apart?

72. Can you place two equal sectors so that one corresponding radius of each sector may be in one line, and so that their angles may point the same way?

[1] Of course it means two lines in the same plane.

73. Upon the same side of the same line, place two angles that shall be equal to each other, and let each angle face the same way.

When two circles have the same centre, they are called concentric circles.

74. Make three concentric circles.

When two circles have not the same centre, and one of them is within the other, they are called eccentric circles.

75. Make two eccentric circles.

76. Draw one line parallel to another line, and let it pass through a given point.

All figures that have four sides take the name of quadrilaterals.[1]

Of quadrilaterals there are six varieties, consisting of quadrilaterals that have their opposite sides parallel, which are called parallelograms; and quadrilaterals that have not their sides parallel, which are called trapeziums.

Of parallelograms there are four kinds: parallelograms, which have all the sides equal, and all the angles equal, called squares. Paral-

[1] Figures of four sides are also called quadrangles.

lelograms which have the sides equal, but the angles not all equal, called rhombuses. Parallelograms which have all their angles equal, but their sides not all equal, called rectangles; and parallelograms which have neither the sides all equal, nor the angles all equal, called rhomboids.

Of trapeziums there are two kinds: quadrilaterals, that have two only of the sides parallel, called trapezoids; and quadrilaterals that have no two sides parallel, which take the name of trapeziums.

77. Give a sketch of a square, of a rhombus, of a rectangle, of a rhomboid, of a trapezoid, and of a trapezium.

The line that joins the opposite angles of a quadrilateral is called a diagonal.

78. Show that each variety of quadrilateral has two diagonals, and say in which kind the diagonals can be of equal lengths, and in which they cannot

In geometry, one figure is said to be placed in another, when the inner figure is wholly within the outer, and at the same time touches the outer in as many points as the respective forms of the two figures will admit.

79. Describe a circle that shall have a diameter of 1½ inch, and place a square in it.

80. Can you make a rhombus?

When a rhombus has its obtuse angles twice the size of those which are acute, it is called a regular rhombus.

81. Can you make a regular rhombus?

82. Can you make a rectangle?[1]

83. Can you make a rhomboid?

84. Can you make a trapezoid?

85. Can you make a trapezium?

When a geometrical figure has more than four sides, it takes the name of polygon, which means many-angled; and when a polygon has all its sides equal, and all its angles equal, it is called a regular polygon.

A polygon that has five sides is called a pentagon.

[1] Rectangles are sometimes called oblongs, and sometimes long squares.

A polygon that has six sides is called a hexagon.

A polygon that has seven sides is called a heptagon.

A polygon that has eight sides is called an octagon.

A polygon that has nine sides is called a nonagon.

A polygon that has ten sides is called a decagon.

A polygon that has eleven sides is called an undecagon.

A polygon that has twelve sides is called a dodecagon.

You have made a sector with a reëntrant angle.

86. Of how few lines can you make a figure with a reëntrant angle?

87. Of how few sides can you make a figure with two reëntrant angles?

88. Of how few sides can you construct a figure with three reëntrant angles?

89. Show how many equilateral trian-

gles may be placed around one point to touch it.

90. Can you divide a circle into six equal sectors?

A sector that contains a sixth part of a circle is called a sextant.

91. Make a sextant, and write upon it its name.

92. Construct an equilateral triangle, and write in each angle the number of degrees it contains.

93. Can you place a circle in a semi-circle?

94. Can you place a hexagon in a circle?

95. Can you divide a circle into eight equal sectors?

A sector that contains the eighth part of a circle is called an octant.

96. Make an octant, and in it write its name, and underneath state the number of degrees that the angle of an octant contains.

97. Make a regular octagon in a circle.

That point in a square which is equally distant from the sides of that square, and also equally distant from the angular points of that square, is called the centre of that square?

98. Draw a line an inch and a half long, and erect a square upon it, and find the centre of it.

99. Can you place a circle in a square?

100. Place three circles so that the circumference of each may rest upon the centres of the other two, and find the centre of the curvilinear figure, which is common to all three circles.

That point in an equilateral triangle which is equally distant from each side of the triangle, and equally distant from each of the angular points of the triangle, is called the centre of the triangle.

101. Can you make an equilateral triangle whose sides shall be two inches, and find the centre of it?

102. Can you place a circle in an equilateral triangle?

103. Can you divide an equilateral triangle into six parts that shall be equal and similar?

104. Can you divide an equilateral triangle into three equal and similar parts?

105. What is the greatest number of angles that can be made with four lines?

106. Make a hexagon, and place a trigon on the outside of each of its boundaries, and say what the figure reminds you of.

107. Can you, any more ways than one, divide a hexagon into two figures that shall be equal to each other, and similar to each other?

108. Can you divide a circle into three equal sectors?

109. Can you fit an equilateral triangle in a circle?

110. Draw two lines cutting each other, and show what is meant when it is said that those angles which are vertically opposite are equal to one another.

111. Can you place two squares so that one angle of one square may vertically touch one angle of the other square?

112. Can you place two hexagons so that

one angle of one hexagon may touch vertically one angle of the other?

113. Can you place two octagons so that one angle of one octagon may touch vertically one angle of the other?

You have divided a line into two equal parts.

114. Can you divide a line into four equal parts?

115. Make a scale of inches, and with its assistance make a rectangle whose length shall be 3 and breadth 2 inches.

116. Draw a line, and on it, side by side, construct two right-angled triangles that shall be exactly alike, and whose corresponding sides shall face the same way.

When a line meets a circle in such a direction as just to touch it, and yet on being produced goes by it without entering it, such line is called a tangent to the circle.

117. Describe a circle, and draw a tangent to it.

The tangent to a circle, at a particular point

in the circumference of that circle, is at right angles to a radius drawn to that point. And as every point in the circumference of a circle may have a radius drawn to it, so every point in the circumference of a circle may have a tangent drawn from it.

118. Can you draw a tangent to a circle that shall touch the circumference in a point given?

119. Given a circle, and a tangent to that circle; it is required to find the point in the circumference to which it is a tangent.

120. Given a line, and a point in that line; it is required to find the centre of a circle, having a diameter of one inch, the circumference of which shall touch that line at that point.

121. Show by a figure how many equilateral triangles may be placed around one equilateral triangle to touch it.

122. Divide a square into four equal and similar figures several ways, and give the name to each variety.

123. Can you place two hexagons so that one side of one hexagon may coincide with one side of the other?

124. Can you divide a circle into twelve equal sectors?

125. Can you place two octagons so that one side of one octagon may coincide with one side of the other?

You have divided a sector into two equal sectors, and an angle into two equal angles.

126. Can you divide a sector into four equal sectors, and an angle into four equal angles?

127. Can you make a rhombus, whose long diagonal shall be twice as long as the short one?

128. Can you make a regular dodecagon in a circle?

129. Can you show how many squares may be made to touch at one point?

You recollect that plane figure that has the fewest lines possible for its boundaries.

130. Of how few plane surfaces can you make a solid body?

A body which has four plane, equal, and similar surfaces, is called a tetrahedron.

131. Make a hollow tetrahedron of one piece of cardboard, and show on paper how you arrange the surfaces to fit each other, and give a sketch of the tetrahedron when made.

You know how to fit a square in a circle.

132. Can you fit a square around a circle?

When two triangles have the angles of one respectively equal to the angles of the other, but the sides of the one longer or shorter respectively than the sides of the other, such triangles, though not equal, are said to be similar each to the other. Now you have made two triangles that are equal and similar.

133. Can you make two triangles that shall not be equal, and yet be similar?

134. Make a rhomboid, and divide it several ways into two figures that shall be equal to each other, and similar to each other, and write on each figure its appropriate name.

135. Make two equal and similar rhomboids, and divide one into two equal and similar trian-

gles by means of one diagonal, and the other into two equal and similar triangles by means of the other diagonal.

136. Can you make two triangles that shall be equal to each other, and yet not similar?

137. Can you show that all triangles upon the same base and between the same parallels are equal to one another?

138. Can you place a circle, whose radius is $1\frac{1}{2}$ inch, so that its circumference may touch two points 4 inches asunder?

139. How many squares may be placed around one square to touch it?

140. Divide a rhombus into four equal and similar figures several ways, and write in each figure its proper name.

141. Show how many hexagons may be made to touch one point.

142. Show how many circles may be made to touch one point without overlapping, and compare that number with the number of hexagons, the number of squares, and the number of equilateral triangles.

When a body has six equal and similar surfaces it is called a hexahedron.

143. Make of one piece of card a hollow hexahedron. Show on paper how you arrange the surfaces so as to fold together, and give a sketch of the hexahedron when finished; and say what other names a hexahedron has.

144. Can you make a right-angled triangle, whose base shall be 4 and perpendicular 6?

In a right-angled triangle, the side which faces the right angle is called the hypothenuse.

145. Can you make a right-angled triangle, whose base shall be 4 and hypothenuse 6?

146. Can you make a rectangle, whose length shall be 5 and diagonal 6?

147. Divide a rectangle several ways into four equal and similar figures, and write upon each figure its proper name.

The term vertex means the crown, the top, the zenith; and yet the angle of an isosceles triangle which is contained by the equal sides is called the vertical angle, however such triangle may be placed; and the side opposite to such angle is still called the base, although it may not happen to be the lowermost side.

148. Place in different positions four isosceles triangles, and point out the vertex of each.

149. Construct an isosceles triangle, whose base shall be 1 inch, and each of the equal sides 2 inches, and place on the opposite side of the base another of the same dimensions.

150. Can you invent a method of dividing a circle into four equal and similar parts, having other boundaries rather than the radii?

You have made a square, and placed an equilateral triangle on each of its sides.

151. Can you make an equilateral triangle, and place a square on each of its sides?

152. Can you fit a square inside a circle, and another outside, in such positions with regard to each other as shall show the ratio the inner one has to the outer?

153. Can you divide a hexagon into four equal and similar parts?

154. Can you divide a line into two such

parts that one part shall be three times the length of the other?

155. Can you divide a line into four equal parts, without using more than three circles?

156. Can you make a triangle whose sides shall be 2, 3, and 4 inches?

157. Make a scale having the end division to consist of ten equal parts of a unit of the scale, and with its assistance make a triangle whose sides shall have 25, 18, and 12 parts of that scale.

158. Can you construct a square on a line without using any other radius than the length of that line?

159. Can you make a circle so that the centre may not be marked, and find the centre by geometry?

160. Can you divide an equilateral triangle into four equal and similar parts?

When a body has eight surfaces, whose sides and angles are all respectively equal, it is called an octahedron.

161. Make of one piece of card a hollow octahedron; show how you arrange the surfaces so as to fold together correctly; and give a sketch of the octahedron.

162. Can you divide an angle into four equal angles, without using more than four circles?

163. In how many ways can you divide an equilateral triangle into three parts, that shall be equal to each other, and similar to each other?

164. Given an arc of a circle: it is required to find the centre of the circle of which it is an arc.

165. Can you make a symmetrical trapezoid?

166. Can you make a symmetrical trapezium?

167. Is it possible to make a rhomboid without using more than one circle?

168. Is it possible to make a symmetrical trapezium, using no more than one circle?

169. Can you place a hexagon in an equilat-

eral triangle, so that every other angle of the hexagon may touch the middle of a side of the equilateral triangle?

170. Can you construct a triangle, whose sides shall be 4, 5, and 9 inches?

171. Can you make an octagon, with one side given?

172. Is it possible that any triangle can be of such a form that, when divided in a certain way into two parts equal to each other, such parts shall have a form similar to that of the original triangle?

173. Show what is meant when it is said that triangles on equal bases, in the same line, and having the same vertex, are equal in surface.

174. Can you divide an isosceles triangle into two triangles that shall be equal to each other, but that shall not be similar to each other?

175. Can you divide an equilateral triangle into two figures that shall have equal surfaces, but no similarity in form?

176. Can you fit an equilateral triangle about a circle?

177. Can you divide an equilateral triangle into four triangles, that shall be equal but dissimilar?

178. Group together seven hexagons so that each may touch the adjoining ones vertically at the angles.

179. Make an octagon, and place a square on each of its sides.

180. Can you convert a square into a rhomboid?

181. Can you convert a square into a rhombus?

182. Can you convert a rectangle into a rhomboid?

183. Can you convert a rectangle into a rhombus?

184. Can you divide any triangle into four equal and similar triangles?

185. Can you invent a method of dividing a line into three equal parts?

186. Can you place a hexagon in an equilateral triangle, so that every other side of the hexagon may touch a side of the triangle?

187. Can you divide a line into two such parts that one part may be twice the length of the other?

188. Can you divide a rectangular piece of paper into three equal strips by one cut of a knife or pair of scissors?

189. You have made one triangle similar to another, but not equal; can you make one rectangle similar to another, but not equal?

190. Can you make a square, and place four octagons round it in such a manner that each side of the square may form one side of one of the octagons?

191. Can you make two rhomboids that shall be similar, but not equal?

192. Can you place a circle, whose radius is $1\frac{1}{2}$ inch, so as to touch two points 2 inches asunder?

193. Can you place an octagon in a square,

in such a position that every other side of the octagon may coincide with a side of the square?

194. Fit an equilateral triangle inside a circle, and another outside, in such positions with regard to each other as shall show the ratio the inner one has to the outer.

195. Can you place four octagons in a group to touch at their angles?

196. Can you fit a hexagon outside a circle?

197. Can you place four octagons to meet in one point, and to overlap each other to an equal extent?

198. Can you let fall a perpendicular to a line from a point given above that line?

Those instruments by which an angle can be constructed so as to contain a certain number of degrees, or by which we can measure an angle, and determine how many degrees it contains, as also by which we can make an arc of a circle that shall subtend a certain number of degrees, or can measure an arc and determine how many degrees it subtends, are called protractors.

Protractors commonly extend to 180°; though there are protractors that include the whole circle, that is, which extend to 360°.

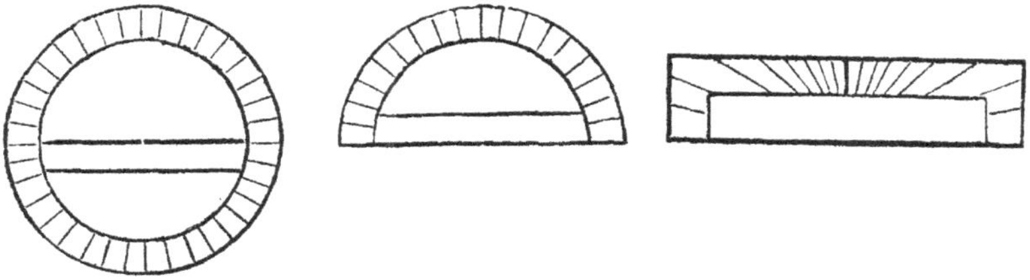

199. Make of a piece of card as accurate a protractor as you can.

200. Make by a protractor an angle of 45°, and prove by geometry whether it is accurate or not.

201. Can you contrive to divide a square into two equal but dissimilar parts?

202. Make with a protractor an angle of 60°, and prove by geometry whether it is correct or not.

203. Make an angle, and determine by the protractor the number of degrees it contains.

204. Make by geometry the arc of a quad-

rant, and determine by the protractor the number of degrees that arc subtends.

205. Show how many hexagons may be made to touch one hexagon at the sides.

That which an angle lacks of a right angle, that is, of 90°, is called its complement.

206. Make a few angles, and say which their complements are.

207. Make an angle of 70°, and measure its complement.

That which an angle lacks of 180° is called its supplement.

208. Make a few angles, and their supplements, and measure them by the protractor.

209. Make by geometry an angle of 30°, and its supplement, and measure by the protractor the correctness of each.

210. Can you make a semicircle equal to a circle?

211. Make a few triangles of different forms, and measure by the protractor the angles of each, and see if you can find a triangle whose angles

added together amount to more than the angles of any other triangle added together.

212. Can you make a pentagon in a circle by means of the protractor?

213. Make of one piece of card a hollow square pyramid, and let the slant height be twice the diagonal of the base. Give a plan of your method, and a sketch of the pyramid, when completed.

214. Can you make a pentagon outside a circle by means of a protractor?

215. Can you, by means of a protractor, make a pentagon without using a circle at all?

It has already been said that the chord of an arc is a line joining the extremities of that arc.

216. With the assistance of a semicircular protractor, can you contrive to place on one line the chords of all the degrees from 1° to 90°? or, in other words, can you make a line of chords?

217. Can you say why the line of chords should not extend as far as 180°?

There is one chord which is equal in length to the radius of the quadrant to which all the chords belong; that is, which is equal to the radius of the line of chords.

218. Say which chord is equal to the radius of the line of chords.

219. Make, by the line of chords, angles of 26°, 32°, 75°, and prove, by the protractor, whether they are correct or not.

220. How, by the line of chords, will you make an obtuse angle, say one of 115°?

221. Can you make, with the assistance of a line of chords, a triangle whose angles at the base shall each be double of the angle at the vertex?

222. Make a triangle, whose sides shall be 21, 15, and 12, and measure its angles by the line of chords and by the protractor.

223. There is one side of a right-angled triangle that is longer than either of the other two. Give its name, and show from such fact that the chord of 45° is longer than half the chord of 90°.

224. Make by the protractor an angle of 90°, and give a figure to show which you consider the most convenient way of holding the protractor, when, to a line, you wish to raise or let fall a perpendicular.

225. Can you make an isosceles triangle, having its base 1, and the sum of the other two sides 3?

226. Can you determine, by means of the scale, the length of the hypothenuse of a right-angled triangle, whose base is 4, and perpendicular 3?

227. Place a hexagon inside a circle, and another outside, in such positions with regard to each other as to show the ratio the inner one has to the outer.

By the area of a figure is meant its superficial contents, as expressed in the terms of any specified system of measures.

In England, the system of linear measures squared is generally [1] used to express areas; as

[1] The terms acres and roods are the exceptions.

square inches, square feet, square yards, square poles, square chains, square miles.

The area of a square whose side is one inch is called a square inch; and a square inch is the unit by a certain number of which the areas of all squares are either expressed or implied.

The area of a square in square inches may be found by multiplying its length in inches by its breadth, or, which is the same thing, its base by its perpendicular height; and as, in the square, the base and perpendicular height are always of equal extent, the area of a square is said to be found by multiplying the base by a number equal to itself, that is, by squaring the base.

228. Make squares whose sides shall represent respectively, 1, 2, 3, 4, 5, etc., inches, and show that their areas shall represent respectively, 1, 4, 9, 16, 25, etc., square inches; that is, shall represent respectively a number of inches that shall be equal to 1^2, 2^2, 3^2, 4^2, 5^2, etc.

229. Make equilateral triangles, whose sides shall represent 1, 2, 3, 4, 5, etc., inches, respec-

tively, and show that their areas (though not actually so much as 1, 4, 9, 16, 25, etc.) are in the *ratio* of 1, 4, 9, 16, 25, etc.; that is, that their areas are in the *ratio* of the squares of their sides.

230. How would you express in general terms the relation existing between the sides and areas of similar figures?

231. Show by a figure that a square yard contains 9 square feet; that is, that the area of a square yard is equal to 9 square feet.

232. Give a figure of half a square yard, and another of half a yard square, and say what relation one bears to the other.

233. Show that the area of a square foot is equal to 144 square inches.

234. Can you show that the squares upon the two sides of a right-angled isosceles triangle are together equal to the square upon the hypothenuse?

Geometricians have demonstrated that a triangle, whose sides are 3, 4, and 5, is a right-angled triangle.

235. Make a triangle, whose sides are 3, 4, and 5; erect a square on each of such sides, and see how any two of the squares are related to the third square.

236. Can you raise a perpendicular to a line, and from the end of it?

237. Can you find other three numbers, besides 3, 4, and 5, such that the squares of the less two numbers shall together be equal to the square of the greater, and show that the triangles they make, so far as the eye can judge, by the assistance of a protractor, are right-angled triangles?

The area of a rectangle, whose base is 4, and perpendicular 3, is 12.

238. Show by a figure that the area of a right-angled triangle, whose base is 4, and perpendicular 3, is half $\overline{4 \times 3}$; i. e., is $\frac{4 \times 3}{2} = \frac{12}{2} = 6$.

A solid bounded by six rectangles, having

only the opposite ones similar, parallel and equal, is called a parallelopiped.

The most common dimensions of the parallelopiped called a building-brick are 9, $4\frac{1}{2}$, and 3 inches.

239. Make of one piece of cardboard a parallelopiped of the same form as a common building-brick;[1] show how you arrange all the sides to fit, and give a sketch of it.

It is now above 2,000 years since geometricians discovered that the square upon the base of any right-angled triangle, together with the square upon the perpendicular, is equal to the square upon the hypothenuse.

You have proved that the squares upon the two sides of a right-angled isosceles triangle are together equal to the square upon the hypothenuse.

240. Can you invent any method of proving to the eye that the squares upon the base and perpendicular of any right-angled triangle what-

[1] When a parallelopiped is long, it takes the name of bar, as a bar of iron.

ever are together equal to the square upon the hypothenuse?

241. Construct a triangle, whose base shall be 12, and the sum of the other two sides 15, and of which one side shall be twice the length of the other.

242. Can you make one square that shall be equal to the sum of two other squares?

243. Can you make a square that shall equal the difference between two squares?

244. Can you make a square that shall equal in surface the sum of three squares.

The angle made by the two lines joining the centre of a polygon with the extremities of one of its sides is called the angle at the centre of the polygon; and the angle made by any two contiguous sides of a polygon is called the angle of the polygon.

245. Make an octagon in a circle, measure by a line of chords the angle at the centre and the angle of the octagon, and prove the correctness of your work by calculation.

A scale having its breadth divided into ten equally long and narrow parallel spaces, cut at equal intervals by lines at right angles to them, with a spare end division subdivided similarly, only at right angles to the other divisions, into ten small rectangles, each of which small rectangles, being provided with a diagonal, is called a diagonal scale.

246. Make a diagonal scale that shall express a number consisting of three digits.

247. With the assistance of a diagonal scale, construct a plan of a rectangular piece of ground, whose length is 556 yards, and breadth 196 yards, and divide it by lines parallel to either end into four equal and similar gardens, and name the area of the whole piece and of each garden.

When a pyramid is divided into two parts by a plane parallel to the base, that part next the base is called a frustum of that pyramid.

248. Make of one piece of card the frustum of a pentagonal pyramid, and let the small end of the frustum contain one-half the surface of that which the greater end contains.

249. Out of a piece of paper, having irregular boundaries to begin with, make a square, using no instruments besides the fingers.

250. Can you show by a figure in what cases the square of $\frac{1}{2}$ is of the same value as $\frac{1}{2}$ of $\frac{1}{2}$, and in what cases the square of $\frac{1}{2}$ is of greater value than $\frac{1}{2}$ of $\frac{1}{2}$?

251. Construct, by a diagonal scale, a triangle whose three sides shall be equal to 791, 489, and 568.

252. Can you show to the eye how much $\frac{1}{2}$ is greater than $\frac{1}{3}$?

253. How many ways can you show of drawing one line parallel to another line, and through a given point?

254. Show by a figure how many square inches there are in a square whose side is $1\frac{1}{2}$ inch, and prove the truth of the result by arithmetic.

255. Show by a figure how many square yards there are in a square pole.

You know how to find the area of a rectangle, and you have changed a rectangle into a rhomboid.

256. How would you find the area of a rhombus?

257. Can you make a right-angled isosceles triangle equal to a square?

258. Can you make a circle half the size of another circle?

259. Can you make ah equilateral triangle double the size of another equilateral triangle?

260. Make of one piece of cardboard a hollow rhombic prism; show how you arrange the sides to fit; and give a sketch of the prism when complete.

261. Make a square, whose length and breadth are 6, and make rectangles, whose lengths and breadths are 7 and 5, 8 and 4, 9 and 3, 10 and 2, and 11 and 1, and show that, though the sums of the sides are all equal, the areas are not all equal.

262. What is the largest rectangle that can be placed in an isosceles triangle?

263. Show by a figure which is greater, and how much, 2 solid inches or 2 inches solid.

If from one extremity of an arc there be a line drawn at right angles to a radius joining that extremity, and produced until it is intercepted by a prolonged radius passing through the other extremity, such line is called the tangent of that arc.

You have given an example of a tangent to a circle.

264. Give an example of a tangent to an arc.

265. Can you draw a tangent to an arc of 90°?

266. Can you contrive to place on one line the tangents to the arcs of all the degrees, from that of one degree to that of about 85°; i. e., can you make a line of tangents?

267. Show which tangent, or rather, the tangent to which arc, is equal to the radius of the line of tangents.

268. Make, by the line of tangents, angles of 20°, 40°, 75°, and 80°.

That solid whose faces are six equal and regular rhombuses is called a regular rhombohedron.

269. Make in card a regular rhombohedron, show how the sides are adjusted to fit, and give a sketch of it when made.

A tangent to the complement of an arc is called the complement tangent, or the co-tangent.

270. Make a few arcs, and their tangents, and their co-tangents.

271. Make an angle, and its tangent, and also its co-tangent.

272. Can you make an angle of 130° by the line of tangents?

273. Can you find out a method of making an angle of 90° by the line of tangents?

274. Measure a few acute angles by the line of tangents.

275. Measure an obtuse angle by the line of tangents.

276. Can you make a rectangle, whose

length is 9, and breadth 4, and divide it into two parts of such a form that, being placed to touch in a certain way, they shall make a square?

277. Show that the area of a trapezium may be found by dividing the trapezium into two triangles by a diagonal, and finding the sum of the areas of such triangles.

278. Make a square, whose side shall be one-third of a foot, and show what part of a foot it contains, and how many square inches.

279. Can you, out of one piece of card, make a truncated tetrahedron, and show how you arrange the sides to fit, and give a sketch of it when made?

280. Can you make a hexagon, whose sides shall all be equal, but whose angles shall not all be equal, and that shall yet be symmetrical?

281. Can you make a right-angled trapezoid equal to a square?

282. Can you make a circle three times as large as another circle?

283. Make by the protractor a nonagon, whose sides shall be half an inch, and measure the angles of the nonagon by the line of tangents.

284. How many dodecagons may be made to touch one dodecagon at the angles?

285. How many dodecagons may be made to touch one dodecagon at the sides?

286. Show by a figure how many bricks of 9 inches by $4\frac{1}{2}$, laid flat, it will take to cover a square yard, and prove it by calculation.

287. Can you determine the number of bricks it would take to cover a floor, 6 yards long and $5\frac{1}{2}$ wide, allowing 50 for breakage?

288. How would you make a square by means of the protractor and a pencil, without a pair of compasses?

289. Can you bisect an angle without using circles or arcs?

290. Construct of one piece of card a hollow truncated cube; show on paper how you arrange

the sides to touch, and give a sketch of the truncated cube when made.

291. Can you make a pentagon, whose side shall be one inch, without using a circle, and without having access to the centre of the pentagon?

292. Can you pass the circumference of a circle through the angular points of a triangle?

293. Show how you would find the area of a reëntrant-angled trapezium.

294. Exhibit to the eye that $\frac{1}{2}+\frac{1}{3}+\frac{1}{6}=1$.

295. Place a circle about a quadrant.

If to one extremity of an arc, not greater than that of a quadrant, there be drawn a radius, and if from the other extremity there be let fall a perpendicular to that radius, such perpendicular is called a sine of that arc.

296. Make a few arcs of circles and their sines.

297. Can you place a circle in a triangle?

298. Can you contrive to place on one line

the sines of all the degrees from 1° to 90°? in other words, can you make a line of sines?

299. Say which of the sines is equal in length to the radius of the line of sines.

300. Given the perpendicular of an equilateral triangle, to construct that equilateral triangle.

When a body has twelve equal and similar surfaces, it is called a dodecahedron.

301. Make of one piece of card a hollow dodecahedron; show on paper how you arrange the surfaces to fit, and give a sketch of the dodecahedron when made.

302. Measure by the line of sines a few acute angles.

303. Can you make an angle of 70° by the line of sines?

The sine of the complement of an arc is called the co-sine of that arc.

304. Show by a figure that the co-sine of the arc of 35° is equal to the sine of 55°.

305. Given alone the distance between the

parallel sides of a regular hexagon, to construct that hexagon.

306. Fit a segment of a circle in a rectangle whose length is 3 and breadth 1.

307. Can you fit a segment of a circle in a rectangle whose length is 3 and breadth 2?

308. Can you place a circle in a quadrant?

309. Give a figure of a symmetrical trapezoid whose parallel sides are 40 and 20, and the perpendicular distance between them 60; measure its angles by the line of sines, and calculate the area.

310. Show by a figure what the area of a rectangle is, whose length is $2\frac{1}{3}$ and breadth $1\frac{1}{8}$, and prove it by calculation.

311. Given, from a line of chords, the chord of 90°, it is required to find the radius of that line of chords.

You have drawn one triangle similar to another, and one rhomboid similar to another; can you draw one trapezium similar to another?

312. Make of one piece of card a hollow em-

bossed tetrahedron ; show how you arrange the surfaces to fit, and give a sketch of it when completed; and say if you can so arrange the surfaces on a plane as to have no reëntrant angles.

313. Can you make one triangle similar to another, and twice the size ?

314. Can you make an irregular polygon similar to another, and twice the size ?

315. Can you make an irregular polygon similar to another, and half the size ?

316. Can you change a square to an obtuse-angled isosceles triangle ?

317. Can you show by a figure how much more $\frac{4}{5}$ is than $\frac{3}{4}$?

318. Can you make an isosceles triangle, each of whose sides shall be half the base ?

319. Can you determine the size of an obtuse angle by the line of sines ?

320. Can you show by a figure that 2 is contained in 3 $1\frac{1}{2}$ time ?

321. Can you show that the sine of an arc is half the chord of double the arc?

322. Take an inch to represent a foot, and make a scale of feet and inches.

323. From the theorem, that triangles on the same base, and between the same parallels, are equal in surface, can you change a trapezium into a triangle?

324. Can you change a triangle into a rectangle?

325. Make of a piece of card a hexahedron, embossed with semi-octahedrons; give a plan of the method by which you arrange the surfaces to fit, and give a sketch of the figure when made.

326. Can you convert a common trapezium into a symmetrical trapezium?

327. Can you construct a square, whose diagonal shall be 3 inches, and find the area of it?

That portion of the radius of an arc which is intercepted between the sine and the extremity of the arc is called the versed sine of that arc.

328. Give an example of the versed sine of an arc.

329. Beginning at a point in a line, can you arrange the versed sines of all the degrees from 1° to 90°? i. e., can you make a line of versed sines?

330. Show when the versed sine of an arc is equal to the sine of the arc.

331. Show when the versed sine of an arc is equal to half the chord of the arc.

332. Say what versed sine is equal to the radius of the quadrant to which the line of versed sines belongs.

333. Given the versed sine of an arc equal to half the radius of that arc, to determine the number of degrees in that arc.

334. Can you reduce a figure of five sides to a triangle and to a rectangle?

When lines or curves, or both, are symmetrically grouped about a point for effect, they take the name of star.

335. Invent and construct as beautiful a star as you can.

When a body has 20 surfaces, whose sides and angles are respectively equal, it is called an icosahedron.

336. Make of one piece of card a hollow icosahedron;[1] represent on paper the method by which you arrange the surfaces to fit, and give a sketch of the icosahedron when made.

337. Describe an arc; let it be less than that of a quadrant, and draw to it the chord, the tangent, and co-tangent, the sine, and co-sine, and the versed sine.

338. Given the sine of an arc, exactly one-fourth of the radius of that arc; it is required, by the protractor, to determine in degrees the length of such arc.

[1] The tetrahedron, the hexahedron, the octahedron, the dodecahedron, and the icosahedron, take the name of regular bodies. These five regular bodies are also called Platonic bodies; and along with these Platonic bodies some place the sphere, as the most regular of all bodies.

339. Given the versed sine of an arc, exactly one-fourth of the radius of that arc; it is required, by the protractor, to determine the degrees in that arc.

340. How would you prove the correctness of a straight-edge, of a parallel ruler, of a set square, of a drawing-board, of a protractor, and of a line of chords?

341. Reduce an irregular hexagon with a reentrant angle to a triangle.

342. Reduce an irregular octagon with two reëntrant angles to a triangle.

It has been agreed upon by arithmeticians that fractions whose denominators are either 10, or some multiple of 10, as $\frac{5}{10}$, $\frac{25}{100}$, $\frac{125}{1000}$, etc., may be expressed without their denominators, by placing a dot at the left hand of the numerator: thus, $\frac{5}{10}$ may be expressed .5; $\frac{25}{100}$ thus, .25; $\frac{125}{1000}$ thus, .125; and $\frac{5}{100}$ thus, .05.

Such expressions are called decimals.

Like other fractions, decimals may be illustrated either by a line and parts of that line, or by a surface and parts of that surface.

343. By dividing a line, supposed to represent a unit of length, illustrate the value of .5, .25, and .125, etc.

344. By means of a square representing a unit of surface, exhibit the value of .5, .25, and .125.

345. Out of an apple, or a turnip, or a potato, cut a cube: call each of its linear dimensions 2, and determine its solid content, and prove by arithmetic.

346. Show by means of a cube, and prove by arithmetic, what the cube of $1\frac{1}{2}$ is.

347. Can you place nine trees in ten rows of three in a row?

348. With 10 divisions of a diagonal scale for its side, construct an equilateral triangle, and call such side 1; and determine the length of its perpendicular to three decimal places, and prove its truth by calculation.

349. Can you calculate the area of an equilateral triangle whose side is 1?

350. Illustrate by geometry the respective values of .9, .99, .999, .9999.

A circle may be supposed to consist of an indefinite number of equal isosceles triangles, having their bases placed along the circumference of the circle, and their vertices all meeting in the centre of the circle. And as the areas of all these triangles added together would be equal to the area of the circle:

To find the area of a circle—multiply the radius which is the perpendicular common to all these imaginary triangles, by the circumference which is the sum of all their bases, and divide the product by 2.

Reckoning the circumference of a circle as $3\frac{1}{7}$ times its diameter:

351. Find the area of a circle whose diameter is 1.

Reckoning the circumference of a circle to be 3.1416 times the diameter:

352. Find the area of a circle whose diameter is 1.

Circles being similar figures, the areas of

circles are to each other as the squares of their radii, their diameters, or their circumferences.

353. Find the area of a circle whose radius is 5, and find the area of another circle whose radius is 7, and see whether their respective quantities agree with the rule.

354. A circular grass-plot has a diameter of 300 feet, and a walk of 3 yards wide round it; find the area of the grass-plot, and also the area of the walk.

355. Can you find the area of a sector whose radial boundaries are each $20\frac{1}{2}$ yards, and whose arc contains 35°?

356. The largest pyramid in the world stands upon a square base, whose side is 700 feet long. The pyramid has four equilateral triangles for its surfaces. Calculate what number of square feet, square yards, and acres, the base of such pyramid stands upon, and the number of square feet on each of its triangular surfaces; calculate also its perpendicular height, and prove its correctness by geometry; give in

card a model of the pyramid; say what solid it is a part of, and give a sketch of the model.

357. There is a rhomboid of such a form that its area may be found by means of one of its sides, and one of its diagonals. Give a plan of it.

358. Can you convert a square whose side is 1 into a rhombus whose long diagonal is twice as much as the short one; and can you find, both by geometry and by calculation, the length of the side of that rhombus?

359. Can you convert an equilateral triangle into an irregular pentagon?

360. Point out upon a tetrahedron two lines that are in the same plane, and two that are not in the same plane.

361. Make of card a truncated octahedron, and give a plan of it, and a sketch of the figure.

362. Show how many cubes may be made to touch at one point.

363. Show by a figure how many cubes may be made to touch one cube.

364. You have calculated the perpendicular height of an equilateral triangle whose side is 1; can you say how far up that perpendicular it is from the base to the centre of the triangle?

A solid formed by revolving a rectangle about one of its sides takes the name of cylinder,[1] and it may be called a circular prism.

365. Can you find the surface of the cylinder whose length is 1, and whose diameter is 1?

A sphere may be formed by revolving a semicircle about the diameter as an axis.

The surface of a sphere whose diameter is 1, is equal to the surface of a cylinder whose diameter is 1 and height 1. Give a figure in illustration of what is meant.

366. Find the surface of a sphere whose diameter is 1, and also the surface of a sphere whose diameter is 2. Compare the two surfaces together, and say whether the ratio the less has to the greater accords with the law,

[1] When a cylinder is long it takes the name of rod, as a rod of iron.

"The areas of similar figures are to each other as the squares of their homologous sides."

367. Can you erect an hexagonal pyramid whose slant sides shall be equilateral triangles?

368. Make a box of strong pasteboard, and let the length be five inches, breadth four, and depth three, and let it have a lid that shall not only cover the box, but have edges clasping it when shut, and hanging over the top of the box three-eighths of an inch.

369. Can you plant 19 trees in 9 rows of 5 in a row?

370. Can you convert a scalene triangle into a symmetrical trapezium?

371. Place a hexagon inside an equilateral triangle, so that three of its sides may touch it, and show the ratio the hexagon bears to the tringle.

372. A philosopher had a window a yard square, and it let in too much light; he blocked up one half of it, and still had a square window a yard high and a yard wide. Say how he did it.

373. Can you divide an equilateral triangle into two equal parts by a line drawn parallel to one of its sides?

374. Given the chord of an arc 50, and the sine of the arc 40; required the versed sine by calculation, and point out on the figure that it is equal to radius minus co-sine.

375. Can you divide a common triangle into two equal parts by a line parallel to one of its sides?

376. Can you divide a triangle into two equal parts by a line from any point in any one of its sides?

377. Show how many solid feet there are in a solid yard.

378. Make an oblique square prism with two rectangular sides and two rhomboidal sides.

379. Make an oblique square prism with all its sides equally rhomboidal.

380. Can you place an equilateral triangle in a square so that one angular point of the equilateral triangle may coincide with one an-

gular point of the square, and the other two angular points of the triangle may touch, at equal distances from the angle of the square, two of the sides of the square?

381. Can you divide a line into 5 equal parts?

382. Can you divide a line into $3\frac{1}{2}$ parts?

383. Can you divide a line as any other line is divided?

384. Can you answer by geometry the question, When three yards of cloth cost 12s., what will five yards cost?

The rule by which areas are found, when the dimensions are given in feet and inches, takes the name of duodecimals; such areas being always expressed in feet, twelfths of feet, or parts, twelfths of parts, or square inches. (*See* Young's "Mensuration.")

Duodecimals are used chiefly by artisans for the purpose of determining the quantity of work they have done, or the quantity of materials they have used.

385. Give a plan of a duodecimal part, that is, of a twelfth of a foot.

386. Give a plan of a duodecimal inch, that is, of one-twelfth of a part, and show that its size is the same as the square inch, although the forms may differ.

387. Give a plan that shall show the area of a square whose side is 1 ft. 1 in., and prove by duodecimals.

388. Give by a scale of an inch to a foot the plan of a board, 3 ft. 4 in. long, and 2 ft. 2 in. wide, and prove by duodecimals the area.

389. Ascertain by geometry how many inches there are in the diagonal of a square foot, and how many in the diagonal of a cubic foot, and prove by calculation.

390. Can you make an octagon which shall have its alternate sides one-half of the others, and that shall still be symmetrical?

391. Can you place in a pentagon a rhombus that shall touch with its angular points

three of the sides of the pentagon and one of its angles?

392. Let there be two rectangles of different magnitudes, but similar in form; it is required to determine the size of another similar one that shall equal their sum.

393. Given two triangles dissimilar and unequal, can you make a triangle equal to their sum?

394. Can you make a triangle equal to the difference of two triangles?

395. Can you make a rectangle equal to the difference of two rectangles?

396. Can you place three circles of equal radii to touch each other?

397. Place a regular octagon in a square, so that four sides of the octagon may touch the four sides of the square.

398. There is one class of triangles that will divide into two triangles that are both equal and similar; there is another class that will bear dividing into two triangles that are similar,

but not equal; and a third class that may be divided into two that are equal, but not similar. Give an example of each.

399. Can you divide a trapezium into two equal parts by a line drawn from a point in one of the sides?

400. Can you place three circles, whose diameters are 3, 4, and 5, to touch one another?

401. Make of strong cardboard a box open at one end, and large enough to receive a pack of cards, and make a lid that shall slide on that end and go over it three-quarters of an inch.

402. Can you make a square that shall contain three-quarters of another square?

403. Can you place a square in an equilateral triangle?

404. Can you place a square in an isosceles triangle?

405. Can you place a square in a quadrant?

406. Can you place a square in a semicircle?

407. Can you place a square in any triangle?

408. Can you place a square in a pentagon?

409. Determine the form of that rectangle which will bear halving by a line drawn parallel to its shortest side, without altering its form.

410. Show that there is a polygon, the interior of which may, by four lines, be divided into nine figures; one being a square, four reciprocal rectangles, and the remaining four reciprocal triangles.

411. Geometricians have asserted that, when in a circle one chord halves another chord, the rectangle contained by the segments of the halving chord is equal to the square of one halt of the chord which is halved; and that, when one chord in a circle halves another chord at right angles, one half of the halved chord is a mean proportional between the segments of the halving chord. Determine, as nearly as you can, by a scale, whether it is true.

One-half of the sum of any two numbers or any two lines is called the arithmetic mean to those numbers, or to those lines.

412. Show by a figure the arithmetical mean to 3 and 12.

The arithmetic mean has the same distance from the less extreme that the greater extreme has from it.

The square root of the product of two numbers is called the geometric mean to those numbers.

413. Show by a figure the geometric mean to 3 and 12.

The geometric mean has the same ratio to one extreme that the other extreme has to it, thus 3 : 6 : : 6 : 12. This is why it also takes the name of mean proportional.

414. Find the arithmetic mean and the geometric mean to 4 and 9. Say which mean is the greater.

415. Determine by geometry and prove by calculation the side of a square that shall just contain an acre.

416. Extract by geometry the square root of 5, and prove by arithmetic.

The angle which the chord of a segment

makes with the tangent of the segment is called the angle of the segment.

417. Can you determine the angle of a segment of 90°?

418. Can you determine which two lines drawn from the extremities of the chord of a segment so as to meet together in the arc of the segment will make the greatest angle?

419. Can you determine the angle in a quadrantal segment?

420. Can you ascertain the relation existing betwixt the angle of a segment and the angle in a segment?

421. Can you give an instance where the angle in the segment and the angle of the segment are equal?

A line that begins outside a circle, and on being produced enters it, and traverses it until stopped by the other side of it, is called a secant to a circle.

422. Make a few circles and fit a secant to each.

A line drawn from the centre of a circle through one extremity of an arc until intercepted by a tangent drawn from the other extremity is called the secant of that arc.

423. Can you make the secant of the arc of 60°?

Like the word tangent, secant has two meanings, one as applied to a circle, and the other as applied to an arc.

Can you on one line, and beginning at one point in that line, place the secants of all the arcs from 10 to 80°? In other words—

424. Can you make a line of secants?

425. Make and measure a few angles by the line of secants.

426. Say which you consider the most convenient for plotting and for measuring angles, the line of chords, of tangents, of sines, of versed sines, or of secants.

427. Calculate the length of a link of a land-chain, and give on paper an exact drawing of such a link.

You have determined how far up the perpendicular of an equilateral triangle the centre is.

428. What ratio have the two parts of an equilateral triangle which are made by a line drawn through the centre of the triangle parallel to the base?

429. Suppose the side of a hexagon to be 1, it is required to determine the sides of a rectangle that shall exactly inclose it, and to find the area of the hexagon and the area of the rectangle, and the ratio between them.

430. Make a figure that shall be equal to one formed by three squares placed at an angle, thus ; and say whether it is possible to divide such figure into four equal and similar parts.

A right-angled triangle made to revolve about one of the sides containing the right angle forms a body called a cone, which may be very properly called a circular pyramid.

431. Make in paper a hollow cone, and give a plan of your method.

When a cone is cut by a plane at right angles to the axis, the section produced is a circle.

When a cone is cut by a plane that makes with the axis an angle that is less than a right angle, but not so small an angle as the angle which the side of the cone makes with it, such section is an ellipse.

A section of a cone making an angle with the axis equal to that which the side makes is a parabola.

A section of a cone which makes, with the axis, an angle that is less than that which the side makes is an hyperbola.

A section of a cone with which the axis coincides is an isosceles triangle.

432. Cut from an apple or a turnip as accurate a cone as you can, and give a specimen of each of the five conic sections.

433. Give a sketch of a builder's trammel, and make an ellipse with a trammel; and show that you can, on the principle on which it acts, make an ellipse without one.

The long diameter of an ellipse is called the axis major, and the short one the axis minor, and the distance of either of the foci of the ellipse from its centre is called the eccentricity of the ellipse.

434. Give a figure to show what is meant.

Two lines drawn from the foci of an ellipse to a point in the circumference make equal angles with a tangent to the ellipse at that point.

435. Can you at a particular point in the circumference of an ellipse draw a tangent to that circumference?

With a pair of compasses, and using different-sized circles, how nearly can you imitate an ellipse? In other words—

436. How would you make an oval?

437. Can you, out of a circular piece of mahogany, and without any loss, make the tops of two oval stools, with an opening to lift it by, in the middle of each?

The solid formed by revolving on its minor axis a semi-ellipse is called an oblate spheroid.

438. Show how an oblate spheroid is formed, and say what the oblate spheroid reminds you of.

The solid formed by revolving on its axis major one-half an ellipse is called a prolate spheroid.

439. Show how a prolate spheroid is formed, and say what it reminds you of.

440. Supposing a room to be built in the form of a prolate spheroid, and a person to speak from one focus, show where his voice would be reflected.

441. Would there be the same effect produced in a room built in the form of an oblate spheroid.

Provided no notice is taken of the resistance of the air, a stone thrown horizontally from the top of a tower, at a velocity of 48 ft. in a second, and subject to the incessant action of the earth, which from nothing induces it to fall by a uniformly-increasing velocity through about 16 ft. in the first second, 48 ft. in the second second, 80 ft. in the third second, 112 ft. in the fourth second, and so on, makes in its progress a kind of curve. Now, the terms of the series 16, 48, 80, 112, 144, etc., increase in a certain ratio; and, if 16 be called 1, 48 will be 3, 80 will be 5, 112 will be 7, and 144 will be 9, etc. These distances may then be expressed as falling distances, thus, 1, 3, 5, 7, 9, etc. And, keeping in mind that the horizontal velocity remains uniform, that is 48 ft., i. e., 3×16 ft. in a second, we have two kinds of dimensions at right angles to each other, from which to make the curve. This curve is called a parabola.

442. Can you construct a parabola?

When these distances, instead of being written down as the separate result of each second's action, are successively added to show the combined results, we have for—

	Falling Distances.		
1 second..........................	1	=	1^2
2 seconds, 1+3=....................	4	=	2^2
3 seconds, 1+3+5=..................	9	=	3^2
4 seconds, 1+3+5+7=................	16	=	4^2
5 seconds, 1+3+5+7+9=..............	25	=	5^2

From this it will be seen that the distance fallen is as the square of the time, i. e., in 6 seconds the distance fallen will be $6^2 \times 16$ ft. $=$ 36×16 ft. $= 576$ ft.

443. Required the distance a stone falls in half a second.

444. Required the distance a stone falls in $2\frac{1}{4}$ seconds.

445. Can you show that there are two kinds of quadrilaterals in which the diagonals must be equal, two kinds where they may be equal, and two kinds where they cannot be equal?

446. Can you make in card a tetrahedron whose four surfaces shall be unlike in form?

9

THE END.

Made in the USA
Monee, IL
08 January 2026

41284523R00059